Why Your Gasoline Prices Are High

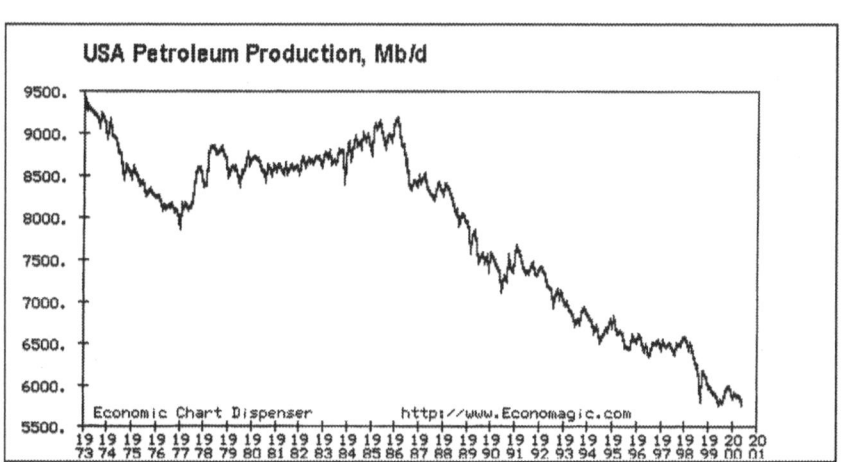

Why Your Gasoline Prices Are High

How you can help to lower them

Seldon B. Graham, Jr.

iUniverse, Inc.
New York Lincoln Shanghai

Why Your Gasoline Prices Are High
How you can help to lower them

iUniverse books may be ordered through booksellers or by contacting:

iUniverse
2021 Pine Lake Road, Suite 100
Lincoln, NE 68512
www.iuniverse.com
1-800-Authors (1-800-288-4677)

ISBN-13: 978-0-595-36940-9 (pbk)
ISBN-13: 978-0-595-81349-0 (ebk)
ISBN-10: 0-595-36940-5 (pbk)
ISBN-10: 0-595-81349-6 (ebk)

Printed in the United States of America

This book is dedicated to the more than
half million loyal, hard-working United States oil and
gas workers who have lost their jobs since 1981.

Contents

Introduction

Every American can express his or her opinion on future gasoline prices and write a book on their opinion. So, why is this book unique?

Few Americans have had 50 years of experience in the oil industry as I have had. Few Americans are both a petroleum engineer and an oil and gas attorney as I am. Only one American founded the United States Oil Association as I did. From this half century of experience in the oil industry, I have assembled *facts* for this book. I have tried hard not to express any opinions.

As a petroleum engineer and expert witness in hearings, and later as an attorney cross-examining petroleum engineering expert witnesses in hearings, I became convinced that a so-called "expert opinion" is only as good as the facts supporting that opinion. Facts are critical. Facts often speak for themselves with little or no opinion necessary to interpret those facts. This book is a book of facts—from an eye witness to oil-related facts during 50 years of experience in the USA Oil industry.

An opinion not supported by any facts is called *ipse dixit* in the legal profession. There are way too many of that kind of opinion being expressed on oil and gasoline. Facts with little or no opinion was the goal of this book.

This book has another asset for the reader. I am a West Point graduate who always strives "to choose the harder right instead of the easier wrong, and never to be content with a half truth when the whole can be won." [from the Cadet Prayer] As Benjamin Franklin said, "Half the truth is often a great lie." I have tried hard to make this book true and honest.

As a graduate of the Army War College and the Air Force Air War College, I know how oil affects national security, particularly as to the mobility of our armed forces. Oil is vital to US national security.

The price of gasoline *is* a political issue. I am too old to expect any profit or benefit from any change in any political party's oil policy. Thus, I have no political agenda other than wanting all political parties to do the right thing. I would like my grandchildren to have the benefit of a bi-partisan 100% Pro-USA Oil policy, something our country has not had in the last quarter century.

I have attempted to be both fair and neutral. In doing so, I made no mention of the name of any political party. However, I must present all relevant *facts* and such facts sometimes relate to a particular politician or a particular political party. Any reasonably intelligent reader will be able to relate such fact with a political party.

This book is about villainization, hate and demonization. My book is not to villainize the villainizers, hate the haters, or to demonize the demonizers. Let bygones be bygones. Forgive and forget. Up to this point, USA Oil Haters—and there are many—may be excused because he or she has been exposed only to repeated lies about USA Oil the likes of which would have pleased Hitler's Minister of Propaganda, the late Joseph Goebbels. Goebbels' philosophy: If you tell a lie often enough, people will believe it.

This book now provides the truth. No longer can "Hate USA Oil" be tolerated. Such actions are un-American, counterproductive, and must stop immediately.

Chapter 1

Do Not Believe Everything You Read, See and Hear about Oil in the Media

I have just read the lead article on the front page of the Austin American-Statesman newspaper this morning. The second sentence in the article says this:

"And gas prices are headed even higher because the cost of crude oil on Monday rose 2.6 percent to a new high, near $64 a barrel."

That statement is wrong, Wrong, WRONG. To repeat Benjamin Franklin, "Half the truth is often a great lie."

I looked to find the author of that statement, hoping to determine his or her expertise. I did not find the author because the article was "from staff and wire reports." Intentionally or mercifully, the author is anonymous.

This current "news" is so typical of what Americans read, see and hear about crude oil from the media. The

American-Statesman is simply following other media like sheep.

First of all, gas is a substance which is often burned in household stoves, furnaces and dryers, correctly called natural gas. The liquid one pours into the tank of an automobile is called gasoline. Please note up front that I use proper nomenclature, but the anonymous writer of the statement did not. So, slang for gasoline *is* "gas." But, did the writer even know the correct name for "gas"? We will never know.

But, that is not what is wrong with the statement. Actually, the first part of the statement seems to agree with the title of this book. The reason given is what is wrong.

It was NOT the <u>cost</u> of crude oil which rose on Monday. It was the <u>price</u> of crude oil on the <u>futures</u> market which rose on Monday.

This is an extremely important distinction. The futures market is for speculation, an euphemism for gambling. It is somewhat analogous to betting on the outcome of the Super Bowl when you don't even know which teams will play. The oil futures market speculates on oil prices 6 months from now. That would be well after next year's Super Bowl.

It is NOT the price that an oilman in the USA got for crude oil on Monday. They may wish that they had gotten that price, but they did not. In the USA, the price of crude oil on Monday was the "posted" price of each purchaser.

Each barrel of crude oil is not created equally. Some oil is more dense than other oil. Some oil contains sulphur. The "posted" price of oil is determined mainly by the gravity (density) and the sulphur content.

The American Petroleum Institute (API) has a measurement for the gravity of oil. Oil is measured in degrees API. The lower the degrees API, such as 23° API, the heavier the oil. The higher the degrees API, such as 43° API, the lighter the oil.

Sulphur content in the oil will reduce the price because it will require an extra step to remove the sulphur. Oil fields with a high sulphur content are known for their rotten egg smell. Thus, the term for high sulphur content crude oil is sour crude oil.

On Monday, an oilman in Wyoming would have probably considered himself lucky to get $38 per barrel for his sour crude oil. Likewise, a Mississippi oilman might have gotten $47 per barrel and a West Texas oilman might have gotten $50 per barrel.

When your business is producing oil, the oil price on the futures market can be a long way from the price you are actually paid. This is a concept lost on the American public.

For those who may want to check this out for themselves, I would recommend the following URL to start with an overview:

< www.oilnergy.com/pr12stat.htm >

When I visited that site last, it had average posted prices for one purchaser for various average API gravity oil from various fields in twelve different states. These averages do not seem to be weighted averages. The point is that even within these different averages, the actual posted prices—*far below the futures price*—may spread over quite different oil prices from relatively high [everything is relative] to very, very low crude oil prices.

The general public is totally unaware that USA Oil (by that term I mean those who produce oil from beneath the USA) did NOT get nearly $64 per barrel for crude oil on Monday. The main stream media is not going to tell the public these important things.

Speculation on oil futures not only becomes "news," that "news" breeds its own baseless brand of speculation and concern. Monday's price of oil on the futures market may or may not be the actual price of oil 6 months from Monday. The actual price will depend upon many factors occurring during the 6 month interval.

The media has time restraints. A writer or reporter does not have years and may not have days in which to research the subject. This is not only true about crude oil prices, it holds true for other stories in the media about oil.

The point to remember from this chapter: USA Oil does NOT get the media's oil price. Do not believe everything you read, see and hear about oil in the media.

Chapter 2

USA Oil has been the Villain for years

Remember the popular CBS television show "Dallas" which ran from 1978 to 1991? Larry Hagman played J.R. Ewing, the greedy, unethical, scheming ultra-rich oil man who used "booze, broads and booty" to make oil deals. That outstanding portrayal of the consummate villain by Hagman left an indelible impression in the minds of Americans. J.R. Ewing became the villain everyone loved to hate. Oil—about which the American public was not knowledgeable—became virtually synonymous with villain.

Had the designated villain been a minority, or had a certain race, creed, religion or nationality, that hit television program probably would not have lasted two episodes because of protests. But, if anyone at all protested the fact that an American oil man was the villain, such protest was ignored. Significantly, CBS did not get sued because the villain in their television series was an American oil man.

Oil men became the perfect villains. No one would protest. No one would sue. If Hollywood should need a villain, Hollywood could safely make that villain an oil man. That would not only be a safe move but it would also be a popular if not expected move because of the "Ewing Oil" implant in the public's brain.

In "Pelican Brief," popular author John Grisham starts with the assassination of two United States Supreme Court Justices. Next, the killing of innocent victims begins. Who is doing this unbelievable evil? Corrupt politicians. What is causing this corruption and unbeliev-able evil? A proposed oil drilling venture—very believ-able to the reading public. The proposed oil drilling might even injure a pelican. Once again, the American oil indus-try became the villain.

In the 1970s, the term "Oil's Seven Sisters" became quite popular to describe the largest oil companies operat-ing in the USA. There was never a consensus as to the members of this so-called sisterhood since ranking could be based on oil production, income, reserves, net worth or many other factors. At the beginning of the 1970s, major oil companies operating in the USA were Humble Oil & Refining Company, Royal Dutch Shell, Texaco, Gulf, Mobil, Sohio, Socal, Amoco, Marathon, Sun, Phillips and Arco which had just purchased Sinclair.

"Seven Sisters" was an unflattering term connoting a very close, blood-is-thicker-than-water, personal relation-ship, perhaps even nepotism, which simply did not exist. Competition among major oil companies was unabated.

In the past, I have been an employee of three of the major oil companies named above and have had close contacts with many of the others during protests of applications and in unit and joint operations. At no time, with the exceptions I will discuss, during my 50 years in the USA oil industry, did I ever witness corporate illegal or unethical conduct by a major oil company.

In fact, illegal or unethical conduct by an employee simply was not tolerated. Corporate decisions, not always sound business practices, were both legal and ethical. Even though oil companies have distinct "personalities," oil companies are amazingly altruistic. Many times, I have witnessed an oil company doing the "right thing," even though such action was against its financial interest.

To name but a few examples of doing the right thing, Arco would consistently recommend that the fairest allocation formula be adopted regardless of its effect on Arco. Mobil would consistently recommend that the fairest allocation formula be adopted regardless of its effect on Mobil. Exxon would consistently recommend that the fairest allocation formula be adopted regardless of its effect on Exxon.

Major oil companies try hard to protect the environment. Exxon had an approved spill plan at the time of the *Valdez* spill. Exxon had aircraft loaded and ready on the runway for take-off to execute the approved spill plan. The Coast Guard ordered Exxon not to execute its approved spill plan. The rest is history.

Enron was a corporate exception and an embarrassment. Enron was born in 1985 with the merger of CEO Kenneth Lay's Houston Natural Gas Company and the old Northern Natural Gas Company. Lay morphed Enron from a transporter and supplier of natural gas into what was primarily a trading company.

During its heyday, I saw an advertisement of Enron indicating that it planned to buy and sell water rights futures. Water! Future water! It was at that point that I realized Enron had ceased to be a real oil and gas company. The fall of Enron came later in 2001.

Kenneth Lay is not an oil man *per se*. He is neither a geologist nor an engineer. His only degrees are in economics. Yet, in a period of only six years, he sent the economics of Enron to the poorhouse. Heraclitus had something to say about such a situation, "Character is destiny."

There were good and honest people who worked for Enron before its collapse. They should not be tarnished for having been employees of Enron.

The recent scandal of Shell's inflation of reserves is another exception and an embarrassment to those in the oil industry worldwide. This was not a technical error. The reserves were artificially inflated by management. After writing an ethics article for the magazine of the Society of Petroleum Engineers, I received an e-mail from a Shell engineer working in Syria who was irate that his London management had increased his reserve estimate. He asked for my help. I sent him a copy of the document he asked for. In addition, I offered to give a free course on business ethics to his management in London, provided that he would clear it with his company. My offer, of course, was not accepted.

The USA was founded on the principle of *personal responsibility.* Individuals are responsible for their own actions. If a person commits a crime, that person should be punished—prison, if appropriate. But, allegations against any person should be backed by factual proof, otherwise, no allegations should be made. Along with the principle of personal responsibility comes the principle of *innocent until proven guilty.*

An entire business organization should not be defamed merely because of the indiscretions of a member. An entire industry should not be defamed merely because of the indiscretions of one company. Blame, if blame is to be placed, should always be focused on a particular individual, with ample proof of the allegations being absolutely necessary. That is the only right thing to do under the USA principles of personal responsibility, innocent until proven guilty, and justice and fair play.

It is wrong and un-American to suggest that companies in the oil business are villains. This is particularly heinous when it is done without any proof.

The point to remember from this chapter: USA Oil is NOT the villain.

Chapter 3

Decline

Every oil and gas well declines in production. That is a fundamental concept in the oil and gas industry. Regardless of the particular time in the life of any well, everything is downhill from that point. The maximum producing rate of a well will decline.

Analysis of this decline is the province of a petroleum reservoir engineer. Accuracy of reserves in the particular reservoir being drained by the well depend upon the accuracy of this analysis.

The only thing that can alter this general decline is a change in conditions. Perforating additional reservoirs if such reservoirs are behind the pipe may increase production. Initiating a secondary recovery project such as waterflooding if such project seems feasible may increase production. Removing accumulated paraffin, if paraffin is present in the tubing, may increase production. Nothing is guaranteed to increase production.

The point is that well production will be on the decline unless something is done—requiring a large expenditure of money—to change it. Production may be increased by workover or remedial work or additional recovery program. However, the well production will begin to decline from the new point. Decline is the way of life of a well.

The "Economic Limit" of a well is another fundamental concept. The Economic Limit of a well is the minimum daily producing rate which keeps the well from losing money. Obviously, an oil producer is unwilling to lose money just to produce oil from a declining well. The Economic Limit is the point at which the well will be plugged with cement and abandoned forever. Government regulatory agencies do not allow a well to stand idle. Such agencies require the well to be plugged with cement to prevent pollution.

Four factors determine the Economic Limit of a well. They are: (1) oil price, (2) daily operating costs, (3) tax, and (4) net working interest. The last three are essentially fixed numbers, leaving price as the only variable.

Since every well is on a decline as it approaches the Economic Limit, an increase in price will cause the Economic Limit to be lower thereby allowing the well to produce longer. Higher prices result in a longer producing life and more reserves. However, if a well has been plugged with cement and abandoned forever after it reached an Economic Limit with a lower price, there is absolutely no help from a higher price since that well is gone forever.

In the past, many USA oil wells have been plugged with cement and abandoned forever—which under present prices would be economical to produce. Thus, the

USA has shot itself in the foot by not allowing wells to sit idle when they reach their Economic Limit until rising oil prices will again allow commercial production. The USA has been forced by government regulation to abandon many oil fields with millions of barrels of oil remaining in the ground because wells are not allowed to remain shut in.

The point to remember in this chapter: Every well declines in production

Chapter 4

The Only Way to Find Oil is to Drill a Hole in the Ground

Oil reserves come in many types. Not knowing that fact can and has caused much confusion about remaining oil reserves. There are undeveloped, developed, speculative, possible, probable and proven reserves. The only reliable oil reserves are developed proven oil reserves.

Developed proven oil reserves requires oil production. No oil production, no proven oil reserves.

Oil production comes from a hole (well) in the ground. The ONLY way to discover oil is to drill that hole. There is simply no other way.

The reverse is also true. If you do not drill, you do not find oil.

Seismic exploration may find possible reserves or some other type of reserves, but it cannot find proven reserves. Only a hole drilled in the ground can do that.

The service company which keeps up with drilling is Baker Hughes. Not surprisingly, that company makes drilling bits for rotary rigs. Drilling rigs are called rotary rigs because the drill pipe rotates.

The point to remember in this chapter: If you do not drill a hole in the ground, you do not find any oil.

Chapter 5

Hubbert's Peak

Hubbert's Peak is the reason why many Americans take the position that the United States has run out of oil. In 1956 Shell research geologist M. King Hubbert presented a paper to the API entitled "Nuclear Energy and Fossil Fuels." On page 16 of this paper, Hubbert assumed that US oil reserves were 150 billion barrels of oil. Based on this estimate of 150 billion barrels of oil reserves, Hubbert predicted on page 24 that oil production in the US would peak about 1965. This is the "Hubbert's Peak" prediction that is forgotten or intentionally ignored because many individuals do not want you to hear about it. This "Hubbert's Peak" prediction was WRONG.

On page 24, Hubbert made a second prediction. He said that, in the event the US oil reserves were 200 billion barrels of oil instead of 150 billion, *allowing himself a 50 billion barrel mistake in his initial assumption,* US oil production would peak about 1970.

This second "Hubbert's Peak" prediction is the one embraced by politicians seeking some political gain in having the USA not producing oil and individuals with a vested interest in the USA not producing oil. Such specious support of Hubbert's Second Peak is, of course, disingenuous.

Hubbert started off his prophecy by *assuming* US oil reserves. Figure 21 in Hubbert's paper is "Ultimate US crude oil based on assumed initial reserves of 150 and 200 billion barrels. Herbert uses the word **"assumed"** in describing his US oil reserves.

Assuming US oil reserves can and has led to a wrong conclusion. US energy policy should never be based upon an *assumption*.

"Hubbert's Peak" is a self-fulfilling and self-defeating prophecy. If you think that you have run out of oil, you will stop looking for oil.

"Hubbert's Peak" was based on *assumed* US oil reserves of 150 billion barrels, and did not come true in 1965. "Hubbert's Peak" was an *assumption* proven to be invalid.

The Department of Energy (DOE) shows cumulative USA Oil production to be 189.6 billion barrels in 2002 and 193.1 billion barrels in 2003. Thus, 3.5 billion barrels of oil were produced in the USA in 2003.

YEAR 2003 WAS THE ALL-TIME PEAK OIL PRODUCTION IN THE USA.

Although I had absolutely nothing to do with this peak, may I claim a discoverer's right to name this 2003 peak "Graham's Peak." Year 2003, yet another man-made "peak," should prove conclusively that the "Hubbert's

Peak" assumption was wrong and that those politicians and environmentalists who embrace "Hubbert's Peak" are also wrong.

The USA has not run out of oil. We should be striving for yet another much higher peak year.

According to the DOE, cumulative USA Oil production was 195.1 billion barrels at the end of 2004. It will only be a couple of years before Hubbert's total ultimate oil recovery assumption of 200 billion barrels is reached. This is more evidence that Hubbert was wrong.

The point to remember from this chapter: The "Hubbert's Peak" assumption that the USA has run out of oil is wrong.

Chapter 6

Price Controls

If you want a product to disappear from the marketplace, set price controls on it. This is a lesson we seem to never learn.

The Edict of 301 A.D. by Roman Emperor Diocletian, which set price controls for the benefit of his army, resulted in goods disappearing from the market. In 1584, Antwerp price controls on food doomed the beseiged city by causing food to disappear.

The Continental Congress set price controls of 30 shillings a day as the maximum price at which an Army quartermaster could hire a driver, a wagon, and four horses. The starving and freezing of US soldiers at Valley Forge was not due to the lack of food and clothing in the states. It was caused by the lack of wagon transportation to get food and clothing to Valley Forge, since wagon owners could obtain three to four pounds per day from private merchants. This cause for the suffering at Valley Forge is documented on page 34 of Erna Risch's Quartermaster Support of The Army: A History of the

Corps, 1775–1939, Quartermaster Historian's Office, Washington, D.C. 1962.

In more recent years, former California Governor Gray Davis should have learned this lesson after he set price controls on natural gas. It created a shortage of natural gas in California.

In 1971, President Nixon proclaimed a national emergency, issued an executive order, and set price controls. US oil production came under price controls. The inevitable result of price controls happened. USA Oil began disappearing from the market.

The point to remember from this chapter: Price controls in 1971 caused US oil production to begin disappearing from the market.

Chapter 7

The USA Oil Plateaus of 1967 to 1974 and 1978 to 1980

USA Oil production was over three billion barrels per year during the 8 years from 1967 through 1974 and also during the 3 years from 1978 through 1980. These periods of time should be more properly described as plateaus of USA oil production rather than a particular "peak" in USA Oil production.

According to the API, USA Oil production in billions of barrels was as follows:

Year	MMM bbls.
1967	3.0
1968	3.2
1969	3.2
1970	3.3
1971	3.3
1972	3.3
1973	3.2
1974	3.0

Year	MMM bbls.
1978	3.0
1979	3.1
1980	3.1

From this data, it should be clear to the reader that the USA did not experience a "peak" in crude oil production. It is more of a plateau. In fact, there were two plateaus. Two plateaus strongly condemn the single "peak" theory.

Many other factors were influencing USA Oil production during these two plateau periods. One factor was the self-inflicted price controls of 1971.

The point to remember from this chapter: USA Oil production had a long 8-year plateau followed by a shorter 3-year plateau during the period from 1967 to 1980, both of which were man-made, but not a single "running-out-of-oil peak."

Chapter 8

S. David Freeman's
Energy: The New Era

In 1974, S. David Freeman wrote a book, <u>Energy, The New Era</u>, Walker & Co., 104 Fifth Avenue, New York, NY 10011. At the time this was written, his used book could be purchased for $2.95. That may be much more than the book is worth, since it demonized USA Oil companies without providing any supporting facts.

Freeman noted on page 79 that, "One might describe the situation as a sit-down strike by the producers, at least within the U.S." USA Oil producers were sit-down strikers, according to Freeman. From this mean-spirited accusation came related accusations that producers had shut in wells to wait for prices to rise.

On page 155 he stated, "One does not have to subscribe to a conspiracy theory to observe that a shortage of energy is a situation most favorable to the energy companies." Here was Freeman's accusation that USA Oil favors shortages. This was a contemptible, below-the-belt charge.

On page 177, Freeman makes the true statement that many people consider "energy companies villains." He does not bother to analyze why or if that conclusion is true or false.

His statements become even more defamatory to USA Oil on page 193. On that page Freeman alleges, "An 'energy crisis', properly managed and manipulated, could be a godsend for industry to override the environmentalist opposition." This infers and suggests collusion and conspiracy by USA Oil.

His highly damaging unsupported allegation on page 230 disclosed Freeman's total lack of knowledge about US Oil production. He alleged, "The industry could produce twice as much oil as presently contemplated from the oil fields already discovered." THIS STATEMENT WAS FALSE.

If Freeman had contacted any knowledgeable person in USA Oil before he wrote that statement, he would have discovered that the statement was false. Obviously, Freeman neglected to make any attempt at finding out the truth.

If a producer can produce a well at its maximum rate—and every producing well in the USA can produce at some maximum rate—the producer would have to be really stupid to produce the well at one-half (1/2) its maximum rate so as to receive only one-half (1/2) the income. On its face and without any research, Freeman's statement is clearly *false*. Freeman intended to deceive the public with this *false* statement. This is a reckless disregard for the truth.

This *false* statement supports the fallacy that USA Oil producers shut in wells to wait for prices to rise. As previously stated, government regulations do not allow a well to stand idle. This malicious smear suggested that USA Oil producers are all law-breakers who ignore government regulations, a false and despicable suggestion.

The more oil a producer produces, the more income the producer receives. An oil producer would have to be really stupid to shut in a well which is capable of producing in order to wait for prices to increase.

Freeman's book provided an insight to his position on other subjects. On page 294 and on page 311, he advocated higher taxes to conserve energy. Regarding private property, he advocated on page 337 that we should "shift our preoccupation for possessions to intellectual and human concerns." On page 338, he suggested a major government program on communes. To save the coastal zones, he proposed on page 301 that the federal government itself explore offshore for petroleum.

Freeman served up his plan for USA Oil's *coup de grace* on page 323 as follows:

"But consumers need protection
against the market power of producers
to charge today's high prices for oil
and gas discovered years ago. Unless
these prices are rolled back and held
in line, the petroleum companies will

continue to reap many billions in
windfall profits
[emphasis added] each year."

Clearly, Freeman ignored the simple fact that profits are absolutely necessary for petroleum companies to have an exploration and drilling budget to drill wells to replace reserves *and continue to exist.* In 1974, Freeman planted the seed in his book for the destruction of USA oil.

The point to remember from this chapter: S. David Freeman's ideas can destroy USA Oil and thereby severely damage the USA.

Chapter 9

President Jimmy Carter

When President Carter came to office, he adopted his famous "moral equivalent of war" energy policy and made S. David Freeman his chief energy advisor. Carter ordered the outside lights of the White House be turned off at night to conserve energy.

President Carter's evening television talks to the nation came right out of Freeman's book, <u>Energy, the New Era</u>. USA Oil took a severe whipping from President Carter who repeated many of Freeman's unsupported allegations.

On January 23, 1980, President Carter announced to Congress that the US would defend the Persian Gulf area by military force if necessary. Thus, the official energy policy of the United States abandoned USA Oil and gave the supply to the Organization of Petroleum Exporting Countries (OPEC).

President Carter with the advice of S. David Freeman, his energy advisor, had embraced the highly questionable second "Hubbert Peak" *assumption* that the USA had run out of oil in 1970. As one of my former professors would say, "If you would believe that, you would believe anything."

Three months after this January 1981 announcement by President Carter, Congress passed the so-called Windfall Profits [a term right out of Freeman's book] Tax. This tax, in effect, sent USA Oil's exploration and drilling budgets straight to the government to spend as it pleased; thereby leaving little or no exploration and drilling budgets for USA Oil.

This was a death notice for USA Oil. The easiest way to explain the Windfall Profits Tax is this: If you bought a house in 1946 for $10,000 and sold it in 1981 for $50,000, your windfall profits tax would be $40,000. Of course, house owners do not have to sell but oil producers had to sell. The only windfall was to the federal government.

Many US oil and gas companies went bankrupt because of the Windfall Profits Tax. Many US oil and gas workers were laid off because of the Windfall Profits Tax.

Those USA Oil companies which survived were forced to go overseas to explore and drill in foreign countries. OPEC was pleased that the USA had sent its oil money and its oil expertise to these foreign countries.

Attention anti-war folks: It was President Carter's energy policy which caused the USA to defend Middle East oil. It was President Carter's energy policy which allowed an Iraqi to kill 37 Americans in one day in the *USS Stark* incident in 1987. It was President Carter's energy policy which caused the US to defend Kuwait when it was invaded by Iraq. These facts are ignored by the anti-war crowd.

The point to remember from this chapter: S. David Freeman's ideas, with the help of President Jimmy Carter and Congress, started the destruction of USA Oil in 1981 which began 25 years of national disaster.

Chapter 10

25 Years of National Disaster

Beginning in 1981, the USA has had 25 years of disaster. In 1981, the average number of rotary rigs running in the USA was 3,974. By 1986, the average number of rotary rigs running in the USA was 964, a loss of over 3,000 rigs. If you do not drill a hole in the ground, you do not find oil. If you do not use a rotary rig, it is soon cut up for salvage scrap.

In 1981, there were about 880,000 USA oil and gas workers, according to a graph on the front page of The Wall Street Journal on June 28, 2005. By 1986, there were only about 590,000 USA oil and gas workers, a loss of about 290,000 USA jobs in only 5 years. By 1999, only about 310,000 USA oil and gas workers were left. The USA had lost over a half million jobs.

In 1986, USA Oil production was about 8,500 barrels of oil per day. By 1999, USA Oil production had plunged to about 6,000 barrels of oil per day, a loss of about 2,500 barrels of oil per day, a decline of over 29% in 13 years.

Ironically, the best illustration of this 25 years of national disaster can be found on the USA Petroleum

Production graph on an anti-USA Oil website, hubbert-peak.com, which clings to the false notion that USA Oil production peaked in 1970 and that the USA ran out of oil in 1970.

This USA Oil production graph is shown below. Clearly, USA Oil production has been plunging downward since 1986.

The question which the "Hubbert's Peak in 1970" folks cannot answer is: Why was there an incline, a trend upward, in USA Oil production from 1977 to 1986? Fools ignore such facts when it does not fit their theory.

The point to remember from this chapter: Since 1981, the USA has had 25 years of disaster causing USA Oil production to plunge downward at an alarming rate since 1986.

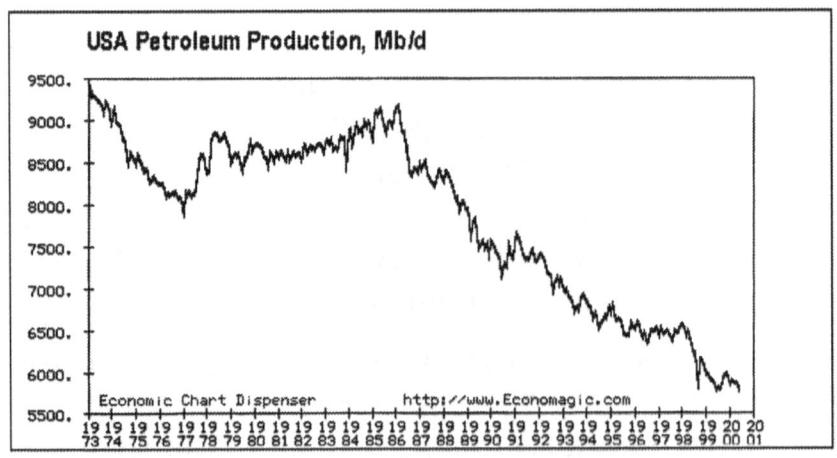

Chapter 11

Hating USA Oil is Exacerbating the 25 Year National Disaster

About half of the American public has grown accustomed to vilifying. maligning and hating USA Oil. USA Oil is blamed for most everything by these individuals who use terms such as Big Oil, Greedy Oil Barons, Villains, Rip-Off Artists, Price-Gougers, Oil Bandits, Big Energy, Record Level Profits, Outrageous Oil Prices, Obscene Profits, Oil's Sweet Deal, Obscene Prices, and still a favorite, Windfall Profits.

It does not take a petroleum scientist to figure out what political entities this is coming from. I shall not mention the name of the political entities in order to protect the guilty. However, for the welfare of the USA, the hate-mongering of fellow Americans must stop.

The targets of this venom against USA Oil companies are mostly foreign. Royal Dutch Shell is headquartered in London. BP is headquartered in England, also. OPEC controls the oil supply to the USA. What remains of USA Oil in the USA are loyal Americans working in USA offices on mostly foreign oil operations. USA Oil was forced to leave the USA by the Carter energy policy which, not incidentally, is causing a lot of wars.

Exxon Mobil is spending a billion dollars in only one field in Russia. This should make Russia and Russian oil workers very happy. Sun disposed of all of its domestic oil production in 1988 to focus on refining and marketing. Shell operates in Russia, Nigeria, Canada and other foreign countries. Arco is in China, Indonesia, the North Sea and other foreign countries. Socal is big in Saudi Arabia. Conoco-Phillips is in Russia, Africa, the Middle East, and Asia Pacific. Marathon is in Russia, Angola, Canada, Guinea and Norway. In other words, so-called "Big Oil" has left for foreign countries.

If the hate-mongers think that the targets of their hate are actually in the USA and that they control the oil supply to the USA, they are both ignorant and wrong. OPEC controls gasoline prices in the USA.

The targets of this venom have also gotten blurry. Sohio merged with British Petroleum (BP) in 1970. Humble Oil & Refining Company, marketing as Esso, changed its name to Exxon in 1973. Socal changed its name to Chevron in 1977. Gulf merged with Chevron in 1984. Amoco merged with BP in 1998. Mobil merged with Exxon in 1998. Texaco merged with Chevron in 2001. Chevron recently purchase Unocal. Today, little is left of oil operations in the USA by major oil companies. Major USA Oil companies who are the refiners and marketers of foreign oil simply cannot do anything about gasoline prices in the USA. OPEC is in control.

Hating USA Oil makes the 25 year disaster more severe. It can do no good. It can do only harm to the USA. USA Oil cannot remedy the 25 year national disaster without an attitude change from the hate-mongers. We are all in the national disaster together.

The point to remember from this chapter: Hating USA Oil is counterproductive and will cause the 25 year national disaster to become even more severe and to continue unabated.

Chapter 12

The Perfect Solution

Twenty-five years of neglect and decline cannot be overcome overnight. It will take much time, much money, and many years. Patience will be needed.

USA Oil must be allowed to make good profits so as to accumulate exploration budgets to find prospects, leasing budgets to buy leases, drilling budgets to drill wells, and development budgets to develop oil fields. USA Oil must be allowed to judge the size of those budgets, not the government or greedy politicians. Much money must be accumulated by USA Oil in order to remedy the problem. This should be a Man-on-the-Moon-type project with the expenditure of USA Oil in the billions of dollars for drilling in the USA.

The Hubbert Peak theory should be trashed. It is self-defeating and invalid. Instead, the Drilling Rig Peak in 1981 of 3,974 rigs should be adopted as the goal to reach. If the rig count is less than that, USA Oil must be allowed to accumulate larger funds to drill more wells.

The manpower peak of 880,000 USA oil and gas workers should also be adopted. If that many Americans are not put to work on correcting the 25 year national disaster, personnel budgets built from profits should be allowed to accumulate to greater levels.

In short, Americans need to have respect for American oil workers at every level in USA Oil companies. They are loyal Americans who deserve respect.

The point to remember from this chapter: It will take a mighty effort, lots of money, and lots of support from all Americans to correct the 25 year national disaster in a reasonable number of years.

Chapter 13

How YOU can Help to Reduce Your Gasoline Prices

You can help to reduce your gasoline prices by not libeling and slandering USA Oil. Even silence would help.

The point to remember from this important chapter: You can help to reduce gasoline prices by not libeling and slandering USA Oil.

Chapter 14

Why Gasoline Prices Will Probably Continue to Rise

Gasoline prices will probably continue to rise because some Americans who have a vested interest in the USA not producing oil will refuse to stop libeling and slandering USA Oil. Stopping the hate campaign against USA Oil might allow proof that the USA has not run out of oil.

This type of American had rather destroy us all than admit a mistake. Such rancor should not be condoned by thinking Americans.

I know of no alternative energy which is currently economical. Thus alternative energy requires taxpayer subsidies. Those seeking government funding of alternative energy benefit if US oil production declines or stops.

The point to remember from this chapter: You cannot afford to tolerate USA Oil hate-mongers.

Chapter 15

USA National Security Needs USA Oil

It has been said that the United States floated to victory in World War II on a sea of oil. That is figuratively true. In spite of shortages of steel and manpower, USA Oil was able to produce the demand which was a lot lower then. At that time, "Willie the Wildcatter" and "Rudy the Roughneck" were as popular as "Rosie the Riveter." How things have changed since.

In 1972, in the Army War College, I wrote a paper relating to Saudi Arabian oil. It was returned to me with a gentle suggestion that I might want to consider writing the paper on another subject which might have a more feasible proposed course of action. I wrote on another subject. In 1972, the military was not interested in Middle East oil.

One of my good friends was Gorman C. Smith. We were at the Army War College together as classmates. Later, Gorman became the Acting Administrator of the Federal Energy Administration, now the Department of Energy. I would kid Gorman by telling him that all he knew about energy was what I taught him during coffee breaks at War College. I did make a trip to the District of Columbia during his tenure to explain to some of his FEA people the methods, techniques and advantages of secondary oil recovery.

In 1988, I wrote a national security article for a newspaper in which I pointed out that it was costing the US taxpayer an estimated $140 per barrel of oil just to defend the oil being shipped through the Strait of Hormuz during peacetime. Foreign oil was not cheap, and we have since learned that it is certainly not cheap to defend.

The Strategic Studies Institute (SSI) of the U.S. Army War College is the Army's organization for the scholarly analysis of national security issues. In 1990 prior to Operation Desert Storm, the Institute issued a special report entitled "Reducing Oil Vulnerability." The War College Commandant kindly furnished me a copy of this report.

I did not agree with this special report. It was written by an Army major who had obviously studied a focused, narrow, non-technical (probably economic) view of oil at some graduate school but had no experience on the subject. The thrust of the special report was that the USA cannot supply its own oil.

I was so concerned that decisions might be made based upon this report, what I called a self-fulfilling prophecy, that I submitted to the SSI my own "Special Report" which objected to and opposed the official one. After providing the background and discussing relevant facts, the first sentence of my recommendation was:

"The United States should adopt
an energy policy which supports
its domestic oil industry rather
than foreign oil importers."

That statement is as true today as it was before the Gulf War. My report ended with this sentence:

"If it is worth dying for in the Near East,
it is worth drilling for in the United States."

My view did not prevail at the time, but no one claimed it was not valid.

In 1994 (during peacetime), the Comptroller of the Department of Defense (DOD) advised me that, for fiscal year 1994, the incremental DOD costs to protect and defend American national security interests in the Persian Gulf region was about $450 million. Taxpayers paid over $1.2 million per day in 1994 for the privilege of purchasing foreign oil. Foreign oil was never cheap oil.

Clearly, the USA dependence on foreign oil has a dominating effect on the Armed Forces of the USA far beyond the uncertainties of an undependable oil supply for its mobility. Our Armed Forces must defend our foreign oil supply.

The point to remember from this chapter: USA National Security needs USA Oil.

Chapter 16

Why USA Oil Remains Undefended

In 1988, I wrote a letter to the magazine of the Society of Petroleum Engineers. It was published in the July 1988 issue as follows:

"Yes, I know that the Society of Petroleum Engineers is an 'international' organization. Does that mean the SPE must sit on its hands and say nothing while the oil and gas industry in the U.S. is destroyed? Is it impossible for the SPE to develop an energy policy for any nation, given a particular set of circumstances?"

"In my opinion, SPE is the only organization uniquely qualified to express an opinion on a U.S. national energy policy. The dead silence of the SPE on such a vital matter may be why there is no U.S. energy policy. It may be why the domestic oil and gas industry is going down the tube."

"Surely, SPE is capable of developing a national energy policy of its own to advocate. For the benefit of the public, isn't it time to get involved, before our industry is completely crushed underfoot?"

"The Energy Focus in the July 1987 JPT (page 774: 'Can the United States Really Afford Low Oil Prices?') was a step in the right direction. But, it would have had much more impact if this one-well example had related, instead, to all wells in the U.S. I doubt if many outside SPE ever saw this Energy Focus. After all, Energy Focus is preaching to the choir when what is needed are missionaries."

"It is time for SPE to raise the hue and cry, to man the ramparts, to be politically alive. SPE needs to take its message—assuming it is capable of developing a message—to the public and to Congress. Politicize or Perish!"

SPE President Lyn Arscott's response was also published. He stated: "Mr. Graham raises a very important issue regarding SPE's policy on matters of national legislative or regulatory policy."

"SPE members hold critical expertise in the technical aspects affecting policy-making in the oil and gas business. SPE members should also be responsible citizens of their countries and be aware of the current legislative and regulatory issues. We need to produce more technical material and analyses so that policy-makers can base their decisions on sound scientific facts. However, there are some restraints."

"First, SPE, Inc. is a 501(c)(3), not-for-profit, tax-exempt association under the IRS rules for the U.S., where we are registered. This means we are registered for scientific and professional purposes. In comparison, a

501(c)(6) association, such as the API, is a trade association whose registration covers lobbying activities."

"Second, many members of our Society enjoy their companies' support for professional development purposes, including their participation in SPE. SPE cannot, by charter, represent the commercial interests of these members' employer companies. We are much better equipped to represent technical issues and the professional interests of our individual members."

"And third, we must be careful as an international society not to favor the political or national interests of one group of members against the interests of another group."

"Given the above pros and cons, I have asked Chairman Fred Wagner of the Technical Information Committee, and his committee members, to prepare a recommendation for the SPE Board of Directors on this subject. I hope this will lead to more activity by the SPE and still allow us to stay within the constraints mentioned above. I look forward to the committee report."

Chairman Fred J. Wagner's response was also provided which said: "As chairman of the SPE Technical Information Committee, I was pleased that you felt the July 1987 'Energy Focus' article was a step in the right direction. We have been striving to prepare 'Energy Focus' articles that inform our readership and promote thoughts such as yours. We ultimately would like to see a wide distribution to other media such as newspapers or trade journals. We are, to some extent, preaching to the choir, but are presently undertaking the challenge to make our communications more effective."

SPE decided to do nothing in the way of developing or supporting any aspect of a national energy policy. Energy Focus disappeared from the pages of its magazine. Yet, as Mr. Arscott said, "SPE members hold critical expertise in the technical aspects affecting policy-making in the oil and gas business." The USA lost a valuable potential asset in establishing a sound energy policy.

The American Petroleum Institute purports to represent domestic oil and gas companies. Because natural gas is part of the API portfolio, natural gas has grown to receive top billing.

The API web site is divided into three parts: (1) Consumer, (2) Professional, and (3) Media. On its Consumer site, little is provided on USA Oil. This site informs the consumer that about 60% of the US oil supply is imported. The answer to the question, "How much oil is left?," says absolutely nothing about USA Oil, only that there are three viewpoints as to the world's oil supply.

The API Professional site has nothing on USA Oil. The oil and natural gas section discusses the work force, clean and safe environment, leaving a lighter footprint, cleaner fuels, keeping America moving, backbone of the economy, and a bright future.

Nothing about USA Oil is readily found on the Media site. This site is devoted to statistics and earnings.

If API considers itself an advocate for USA Oil, and I am not certain that it does, it is an extremely poor advocate. The fact that the *ONLY* information regarding oil which is readily found on the API web site—that about 60% of the US oil supply is imported—shows gross neglect in advocating for USA Oil. Even that fact on imports does not make mention of USA Oil.

No nation-wide organization has been defending USA Oil against critics, and there are legions of critics, in the media, in Congress, and in the public. There simply is no national organization which will stand up to defend USA Oil and provide the truth to the American public.

Of course, this lack of defense has delighted the critics, many of whom seek political gain by attacking "Big Oil." The public is inundated with half truths about oil, in spite of Benjamin Franklin's caveat, "Half the truth is often a great lie." Some propaganda against USA Oil does not even contain half the truth.

Since cheaper USA Oil is the only hope for consumers to have lower gasoline prices, now is the time for consumers to challenge the facts and the motives of those hypercritical of USA Oil. Only the individual consumer can cause a change in public attitude since there is no national organization to help.

The point to remember from this chapter: You must help to lower your own gasoline prices by taking overt actions to defend cheaper USA Oil.

Chapter 17

A Public Service Announcement

(1) The intention to deceive is necessary for a false statement to be a "lie."

(2) Making a false statement against an individual or individuals with the intention to deceive others is a lie.

(3) Public figures are the only individuals about whom you can lie with impunity. Those who lie about public figures are immune from libel and slander lawsuits provided the lying is not done maliciously. [That is a sad state of affairs.]

(4) USA oil and gas workers are not public figures, at least I do not know of one who is.

(5) Laid off USA oil and gas workers make up a very large "class" of individuals.

(6) Lying about USA Oil could have consequences other than causing your gasoline prices to continue to rise.

If you do not lie about USA Oil, you need not remember anything about this chapter. However, if you do, then you have received fair warning.

Chapter 18

Miscellaneous Facts

By now, it should be no surprise to you to learn that S. David Freeman was the Chief Energy Advisor to California Governor Gray Davis during the natural gas shortage debacle. Freeman became CEO of the Hydrogen Car Company. Freeman was a participant with former Enron executive Jeff Skilling—before Skilling was indicted for fraud—in a panel to promote hydrogen cars.

Scott Tinker, Director of the Bureau of Economic Geology at the University of Texas is quoted in The Alcalde, the magazine of the Ex-Students Association, as saying peak oil makes good press but really doesn't matter; it's really an exercise in the unknowable.

Why would anyone be mean to Dr. Seuss? Dr. Seuss (Theodor Seuss Geisel) once was "Big Oil." He was an advertising illustrator for the company which is now Exxon Mobil.

Texas has an anti-littering bumper sticker which says, Don't Mess With Texas. (Translation for those only familiar with the old Army terminology for meals: Don't trash

Texas.) I would like to have an anti-trashing bumper sticker which says, Don't Mess With USA Oil.

You need not remember anything in this chapter, but I will bet that you do.

Recapitulation

1. USA Oil does NOT get the media's oil price. Do not believe everything you read, see and hear about oil in the media.

2. USA Oil is not a villain.

3. Every well declines in production.

4. If you do not drill a hole in the ground, you do not find any oil.

5. The "Hubbert's Peak" *assumption* that the USA has run out of oil is wrong.

6. Price controls in 1971 caused US oil production to begin disappearing from the market.

7. USA Oil production had a long 8-year plateau followed by a shorter 3-year plateau during the period from 1967 to 1980, both of which were man-made, but not a single "running-out-of-oil peak."

8. S. David Freeman's ideas can destroy USA Oil and thereby severely damage the USA.

9. S. David Freeman's ideas, with the help of President Jimmy Carter and Congress, started the destruction

of USA Oil in 1981 which began 25 years of national disaster.

10. Since 1981, the USA has had 25 years of disaster causing USA Oil production to plunge downward at an alarming rate since 1986.

11. Hating USA Oil is counterproductive and will cause the 25 year disaster to become even more severe and to continue unabated.

12. It will take a mighty effort, lots of money, and lots of support from all Americans to correct the 25 year national disaster in a reasonable number of years.

13. You can help to reduce gasoline prices by not libeling and slandering USA Oil.

14. Do not tolerate USA Oil hate-mongers.

15. USA national security needs USA oil.

16. You must help to lower your own gasoline prices by taking overt actions to defend cheaper USA Oil.

Postlude

a/k/a
Parting Shots

I am not a celebrity. I am not a public figure, and I do not desire to be one. I am not a greedy, price-gouging, villainous, fat-cat oil baron—and I am tired of being called one. To repeat a famous movie line, "I'm mad as h___, and I'm not going to take it anymore."

I have a DUTY to my COUNTRY to write this book. It is an HONOR for me to be able to contribute my knowledge and experience to this effort.

I am not seeking fame or fortune by this book. I am trying to pass out information.

I have tried to apply the KISS principle [Keep It Simple, Stupid] so that even my liberal friends who consider themselves intellectuals and Bill O'Reilly will be able to understand it. I could have added many more chapters. e.g. United States Oil Association. But, in attempting to keep the book short, concise and to the point, this is what you get.

I intend to price this book so that every driver in America can afford to buy it. Gasoline price complainers will have *no excuse* for being uninformed.

The current misplaced anger about gasoline prices is analogous to a dying person becoming outraged at the physicians who are trying desperately hard to save the person's life and firing all of them. That is not a good survival plan.

Last but certainly not least, thank you very much for reading this book. My hope is that you will join me in trying to reduce **your** gasoline prices.

Seldon B. Graham, Jr.

978-0-595-36940-9
0-595-36940-5